'Nano in Nature' Grades K- 5

The Magic of the Nano World

is in your pencil

'The Next Generation of Science Standards' has been released by the National Academy of Science

Kindergarten through 5th Grade:

Students in kindergarten through fifth grade begin to develop an understanding of the four disciplinary core ideas: physical sciences; life sciences; earth and space sciences; and engineering, technology, and applications of science. In the earlier grades, students begin by recognizing patterns and formulating answers to questions about the world around them. By the end of fifth grade, students should be able to demonstrate grade-appropriate proficiency in gathering, describing, and using information about the natural and designed world(s).

The performance expectations in elementary school grades, to develop ideas and skills that will allow students to explain more complex phenomena in the four disciplines as they progress to middle school and high school. While the performance expectations shown in kindergarten through fifth grade couple particular practices with specific disciplinary core ideas, instructional decisions should include use of many practices that lead to the performance expectations.

Non-fiction

Nanoscience Education, Workforce Training, and K-12 Resources: Authors: Judith Light Feather, Miguel F. Aznar, CRC Press. ISBN: 978-1-4200-5394-4

Nanoscience and nanomaterials: synthesis, manufacturing and industry impacts: Wei-hong Zhong et al: DESTech Publications, ISBN: 978-1-6059-5013-6. *Chapter 12 – Co-authors: Michael Richey, Boeing Corporation, Judith Light Feather, The NanoTechnology Group Inc., Robert Cormia, Foothill College*

NANOTECHNOLOGY: EHTICAL AND SOCIAL IMPLICATIONS: Edited by Ahmed S. Khan, CRC Press, ISBN: 978-1-4398-5953-7, *Chapter 13: What Are the Social Implications of Our Delay in Teaching Nanoscience Education to K-12 Students in the United States? Contributor: Judith Light Feather*

Nanotechnology: Business Applications and Commercialization: Sherron Sparks, PhD, CRC Press, ISBN:978-1-4398-4521-9, *Chapter 10: Support Organizations, Case Study: The NanoTechnology Group Inc.: A Global Education Consortium, Page 160-167. Contributor: Judith Light Feather*

©Judith Light Feather

This textbook has been designed for and previewed by the NSEE workshop for nano science education in Washington DC, December 2014. http://nseeducation.org/2014/ The Next Generation of Science Standards and the NSF work progressively to encourage teachers understanding and inclusion of nanoscale science and technology in K-12, using new STEM integrated interactive blended curricula.

Cover Photo: *Courtesy of stock images at freedigitalphotos.net*

Table of Contents

Introduction for parents and teachers

This book has been created to introduce the study of nature by combining the geometry shapes your child must learn in math, with the nanoscale size in science. Nano means 'dwarf' in Greek - and has set the world of science on a new path of discovery - that has already changed our world. The working devices on chips in our cell phones, computers, automobiles, planes, ships, and satellites have dimensions down to 30 nanometers or less.

Scientists and researchers focus on aspects of nature as separate functions, looking for results in each experiment to validate a rule or law. Everything is separated into different subject matter and labelled, from the tiniest energy particle to the cosmos.

Children are required to learn the geometry shapes of the Platonic solids, but are not aware that they are the foundation of chemistry, biology, physics, and astronomy . These geometric shapes are replicated by the many atoms in the periodic table, which bond to form everything in the natural world.

Life sciences could be taught as Biomimmicry, since scientists and engineers are really trying to figure out how; a) nature works, b) to copy it, c) how to change DNA and protein structure by manipulating and moving atoms. This all begins at the nanoscale of science.

At this very small size, which is a billionth of a meter, everything is in motion. Structural patterns and the relationships of energy with movement are the foundation of everything in our world.

Scientists still do not understand how the energy becomes mass at size scales even smaller than the nanometer from the tiny world of atoms and energy, or how it all becomes mass...they are spending billions on the CERN particle accelerator to find a Higgs-Boson, that they think is involved in the transition that changes energy into mass.

There are trillions of atoms in our physical bodies that self-direct the production of new cells every seven days, while the neurons in your brain automatically flash to operate the body like a well-oiled machine. The body has a built-in immune system to protect against harm, and manufactures energy for every cell without our conscious instructions.

My intention is to stimulate the natural desire for learning and the creativity in our children. The beauty of nature still holds many secrets that they may discover in the future.

It is my hope that using the new rules of blended learning will enable the young students to watch the recommended videos in classrooms, and use the interactive resources. These visual elements offer a better understanding of the world, and help them 'connect the dots' as they move into the higher grades.

Let's put the "WOW" back in education!

Chapter 1:

The Magic of the Nano-size World!

In 1985, during an experiment to study soot in his laboratory, forms of carbon with 60 atoms were accidentally discovered.   Dr. Richard Smalley, the team leader stated, "I do not want to study smoke." He was really surprised when he found the carbon 60, (C^{60}) molecule, and his team won the Nobel Prize in Chemistry, in 1996.

How does this involve math?

They could not see the shape of the molecule yet, as their tools were not powerful enough, but they knew it had 60 atoms. Based on the work of an 18th century math expert named Leonard Euler, who stated that, "every closed polygon made with hexagons and pentagons, must contain 12 pentagons," C ^{60}'s soccer-ball shape emerged as the smallest possible structure, in which the 12 pentagons, do not touch. In any smaller structure, the pentagons must touch. The new molecule was also, hollow inside. This illustration from NASA in 2010, shows that space is full of these nano-size 'buckyballs', named after the architect, Buckminster Fuller, who created the first model of a geodesic dome, which incorporates this structure.

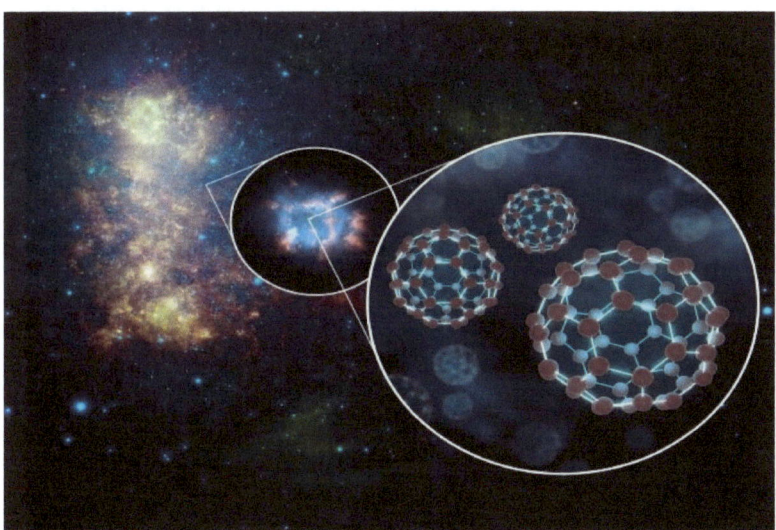

Photo: NASA – 2010. Astronomers have discovered bucket loads of buckyballs in space. They used NASA's Spitzer Space Telescope to find the little carbon spheres throughout our Milky Way galaxy - in the space between stars and around three dying stars. What's more, Spitzer detected buckyballs around a fourth dying star in a nearby galaxy in staggering quantities -- the equivalent in mass to about 15 of our moons.

Geometry is part of the math of nature. While playing soccer, imagine all those buckyballs in space, moving between the stars.

We live on a carbon based planet, in a carbon based universe. As we explore nature at the nanoscale size of science, the geometry shapes learned in math will be found throughout nature, helping you to learn the importance of their movement.

National Standards for geometry K-5 :

http://www.platonicsolids.info/standards.htm

The drawing shows that all the platonic solids in your math class fit within each other inside of a sphere. This will help you to understand how they change shape when they move in nature.

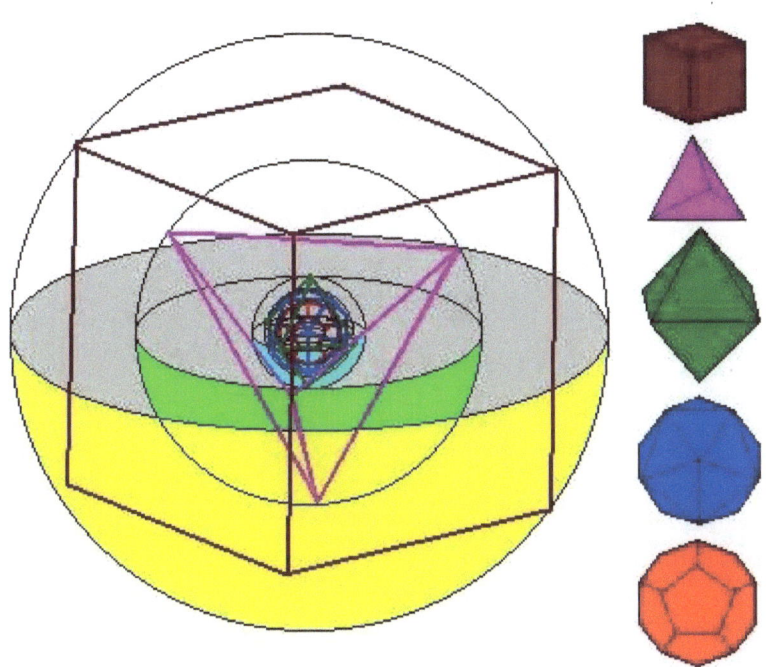

At the nanoscale of science the 'buckyball' made of 60 atoms, as described earlier, is a closed polygon, made with 12 hexagons and 12 pentagons, which opened the window for research into this tiny size of nature. The video from the Vega Trust shows you how to build a buckyball in your classrooms.

To view a video on the Buckyball visit:

http://vega.org.uk/video/programme/163

Nanotubes or fullerenes:

A second video will show, how sheets of graphene in our experiment, can be made into fullerenes, which have 36 carbon atoms shaped as hexagons, and are used in nanotechnology, due to their light weight and strength, when added to materials.

http://vega.org.uk/video/programme/223

Chapter 2
Experiment with a pencil!

The photo below is of a sheet of graphene and shows the pattern of six-sided hexagons, from your math class on geometry.

Sheet of graphene from the graphite in a pencil.

Graphite from a pencil is made of 3 million very

tiny layers of graphene. A single-layer sheet is 1 atom thick, 1 billionth of a meter.

 They are so light in weight that a single-layer sheet -- big enough to cover a whole football field -- would weigh less than one ounce.

Each sheet is made of 6-sided hexagons, with the atoms holding them all together.

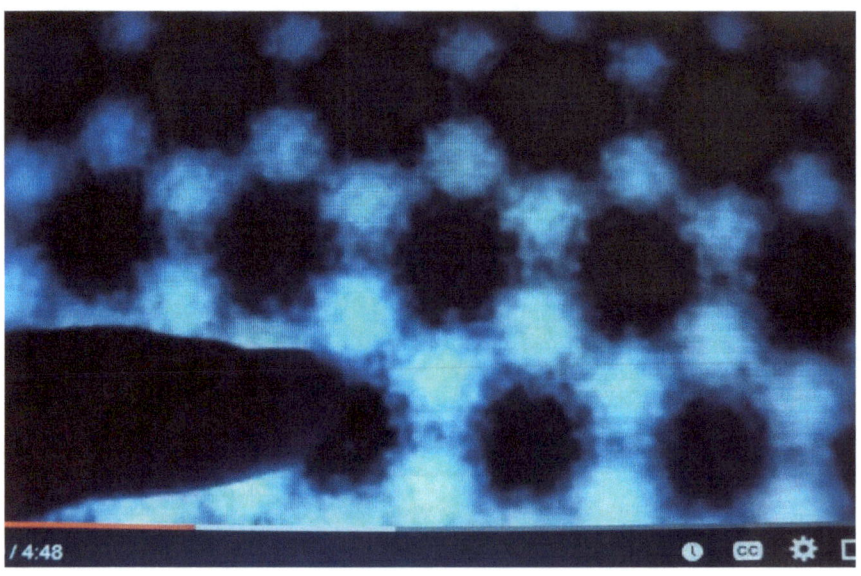

/ 4:48

Atoms (white circles) from special atomic force microscope.

Why is Carbon important?

Carbon is the base for all life forms on earth. We

have so much Carbon (C^6) in our bodies that it is second highest element, and in the universe it is the sixth highest element.

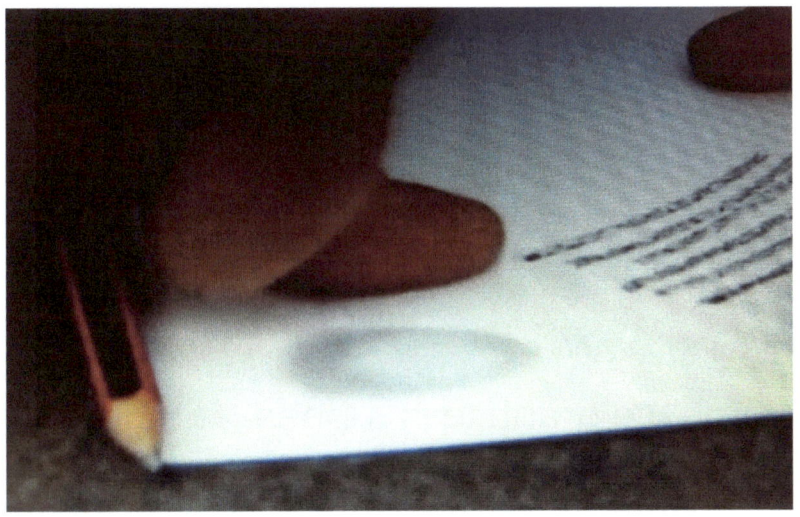

Each thumb print is a layer of graphene.

When writing with a pencil, it leaves multiple layers of soft graphene on the paper as the pencil moves.

When rubbing a thumb on the letters or lines, a layer will attach to the skin. Then make a thumbprint to see each layer.

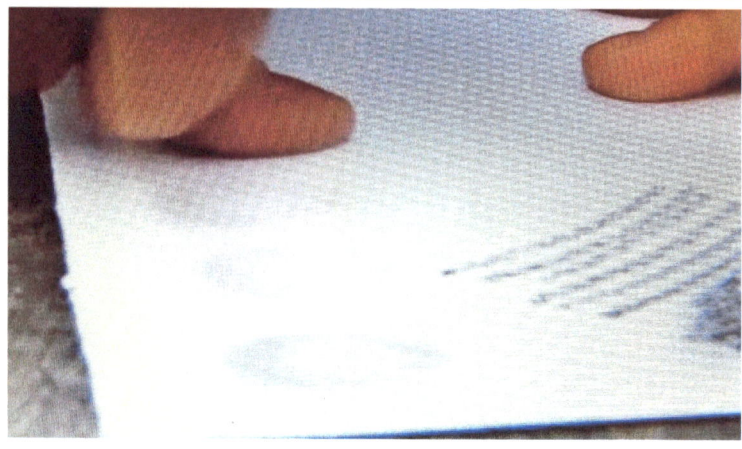

Many thumbprints can be made and each layer is thinner.

Each time the graphite is rubbed, another layer is removed. Since there are millions of layers, it would take a long time to remove all of them.

The individual graphene layers are very strong and bendable. The next part of the experiment demonstrates this feature. You can try it with a piece of scotch tape as shown.

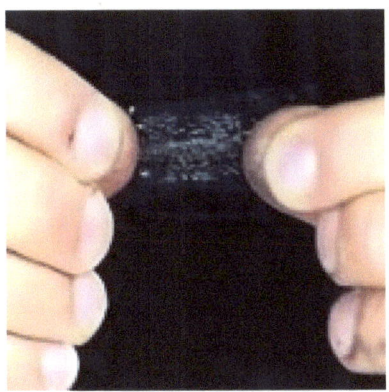

Scotch tape picks up more of the graphene.

Using a piece of scotch tape instead of a thumbprint, shows the strength of the graphene, and the flexibility, if applied to thin plastic. As the tape is opened and closed, the tiny pieces stay on. They will not fall off.

The next part of the experiment will show that electricity flows through graphene from the pencil.

In 2010, a team in the U.K. proved it can conduct even the tiniest amounts of electricity at the nano-size and won a Nobel Prize.

Engineering and new technology from the discovery of graphene!

It is so valuable that flexible roll-up screens will replace rigid computer screens in the future, along with wearable computers in clothing, and plastic solar panels in the bodies of automobiles, that charge the batteries of electric cars.
The photo shows the flexible plastic that will be used in the future for manufacturing.

Prof. Jari Kinaret
Director
The Graphene Flagship

Video of electrical properties in the Experiment with your pencil

The simple experiment shown in the next video, has a 9 volt battery connected to a small LED bulb, with two strips of metal.

Classroom experiment shows the LED bulb light up.

At the top left of the photo, when he goes back and forth with the pencil across the metal, the light comes on. Watch the video at:

http://vega.org.uk/video/programme/326

In October 2010, the Nobel Prize for Physics, was won by Andre Geim and Constantin Novoselov, at Manchester University, for their work on graphene.

The Vega Trust has a wonderful series of programs, and a sub-series at:

http://vega.org.uk/video/subseries/27

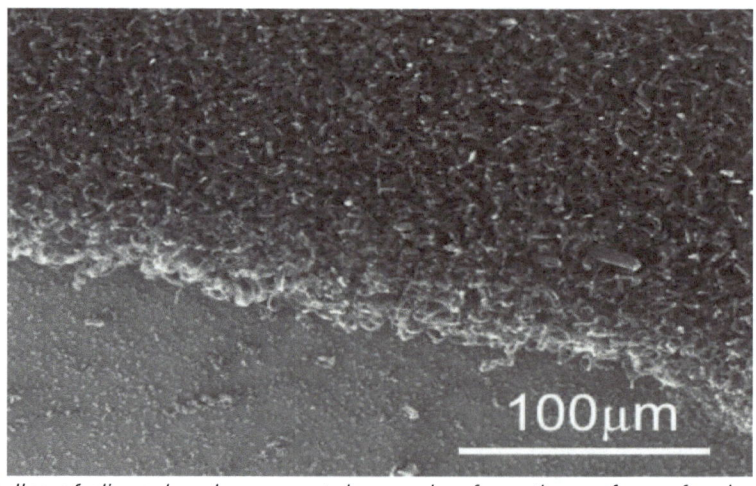

Bundles of aligned carbon nanotubes spring from the surface of a sheet of graphene. The millions of nanotubes shown here are covalently bonded to the graphene, meaning they are essentially a single surface. The material invented at Rice is being used by materials scientists as a cathode for dye-sensitized solar cells. (Credit: N3L Research Group/Rice University)

This discovery will lead to new types of solar cells.

18

Chapter 3:
Energy spirals in nature and in space.

Energy at the nanoscale size down to the planck scale, which is the smallest known size, behaves differently than kinetic energy. It can be a particle or a wave, and can blink in and out of space, and also appear in many places in the universe at the same time. This is known as quantum behavior, which Einstein called "spooky". This energy is in the spirals of galaxies and the spirals of our weather patterns. It has been difficult in the past to show this pattern of movement that involves geometry and math. The study of these patterns and movement of the geometry, as it spins in nature, can seem magical and always includes - you - as the observer. Most of the observations at this size require special cameras, but sometimes we get lucky, and nature allows us to see these spirals as a prism in a sunbeam.

The following photo shows one of these spirals in nature, while photographing a large turtle.

Photo: Fort Worth Nature Preserve,Texas, Laura Zimmerer

The Geometry of sea shells repeats itself without any instructions from man. Many of the patterns on insects, butterflies, turtle shells, and even our vegetables,flowers and trees, grow with the same geometry structure. We call it nature by numbers, repeating over and over again throughout the universe.

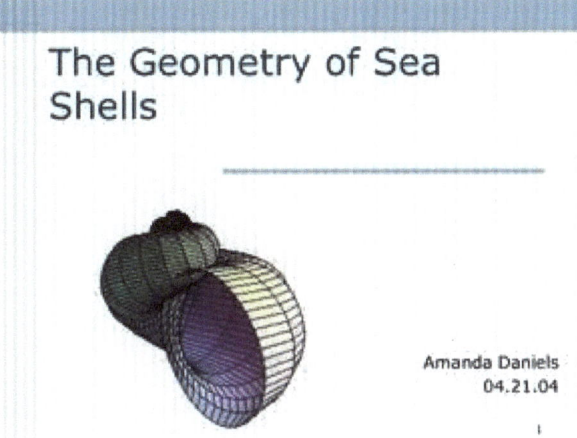

The Geometry of Sea Shells

Amanda Daniels
04.21.04

This geometry of a sea shell, follows the math from the Fibonacci Spiral, which is found in plants, shells and even in space.

This video shows how the spiral works.

Nature by Numbers

https://www.youtube.com/watchv=kkGeOWYOFoA

Fibonacci Fractals

Shows how these numbers scale up from the tiny atom to the universe.

https://www.youtube.com/watch?v=bE2Eil-UfsE

The Torus in space

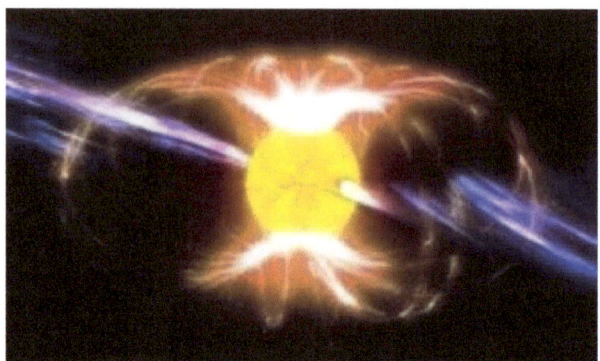

NASA Photo: A magnetar releasing energy from a star, shows the single torus shape.

The following illustration shows how a double torus moves, and releases energy from the vortex points of the star.

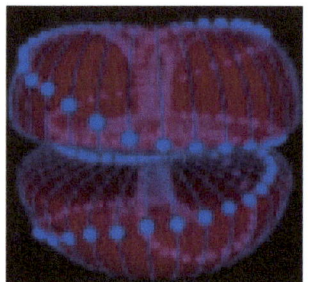

Nassim Harriman, Renaissance Project

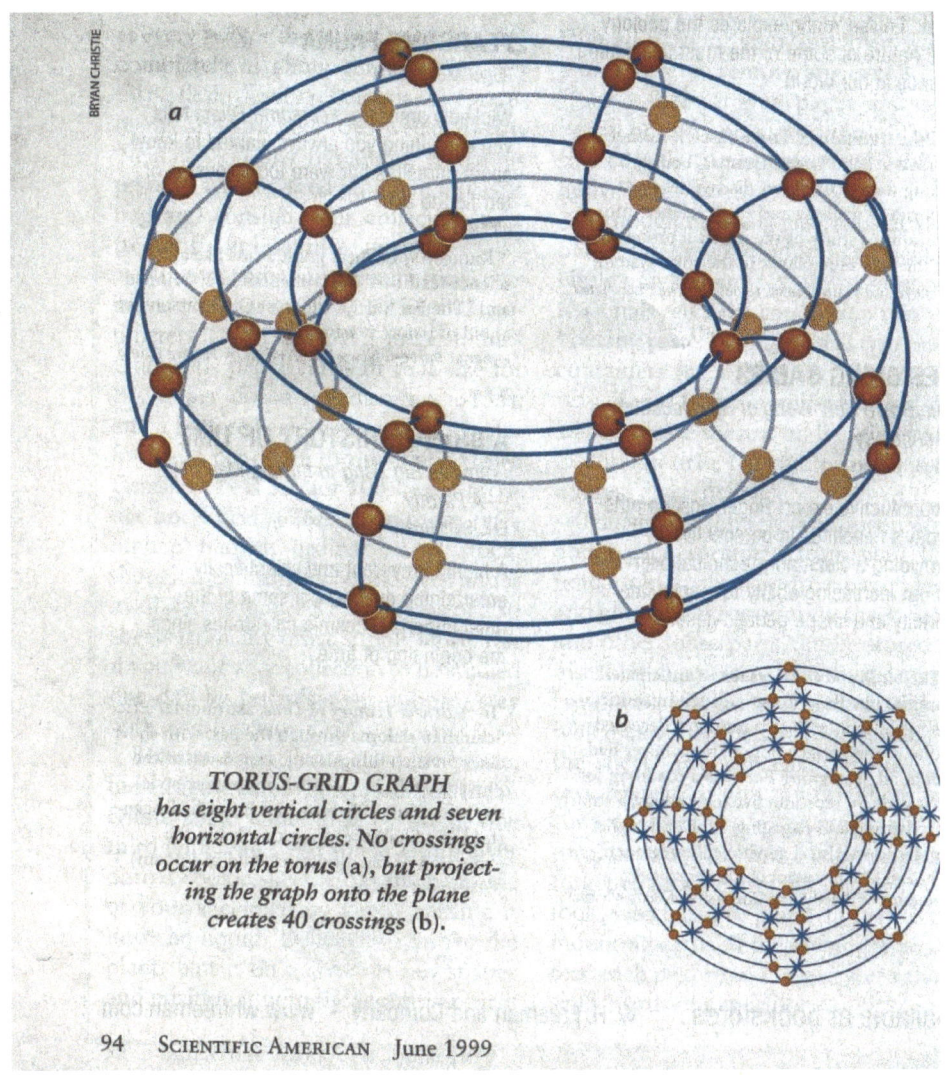

BRYAN CHRISTIE

a

b

TORUS-GRID GRAPH
*has eight vertical circles and seven
horizontal circles. No crossings
occur on the torus (a), but project-
ing the graph onto the plane
creates 40 crossings (b).*

94   SCIENTIFIC AMERICAN   June 1999

Graphic from Scientific American, June 1999
The Torus-Grid Graph has eight vertical circles and seven horizontal

circles and can be any size, from the nanoscale to the cosmos.

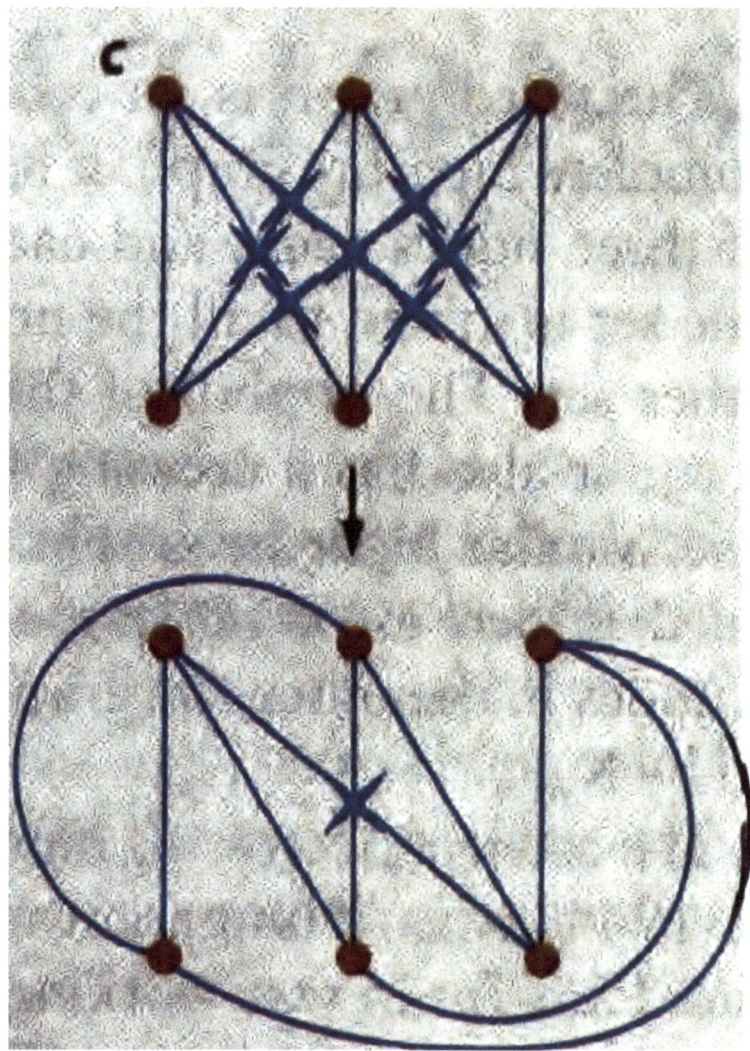

Graphic from Scientific American, June 1999.

The diagram of the pattern in the torus-grid movement shows how six energies at a time can release through a point without ever crossing or touching, as seen in the magnetar photo from NASA.

24

Chapter 4:
What about our bodies?

It might be surprising to learn that we have trillions of these atoms in our bodies. Every cell and molecule has atoms, that are constantly moving and performing tasks to keep our bodies healthy and functioning. It would be hard to imagine all that movement in the cells, and how the DNA makes more cells, without directions. The video shows everything at work.

"Molecular Visualizations of DNA":
https://www.youtube.com/watchv=4PKjF7OumYo

Secret Worlds: The Universe Within - Understanding Size

View the Milky Way at 10 million light years from the Earth. Then move through space towards the

Earth in successive orders of magnitude, until reaching a tall oak tree, just outside the buildings of the National High Magnetic Field Laboratory in Tallahassee, Florida. After that, begin to move from the actual size of a leaf into a microscopic world that reveals leaf cell walls, the cell nucleus, chromatin, DNA and finally, into the subatomic universe of electrons and protons.

Set the controls to stop and view each part, or choose automatic.

Students can view nature from the cosmic size of the universe down to the atoms at the nanoscale of science, gaining a better idea of this tiny size in nature.

http://micro.magnet.fsu.edu/primer/java/scienceopticsu/powersof10/index.html

Chapter 5:
Size Matters - Understanding Nano
Size Matters: The Scale of the Universe 2
down to the Planck scale

Interactive slide control, allows students to view the entire set of known scales, and the wavelengths of each scale.

A wonderful introduction to all ages that teachers can offer in their classrooms. Try it out with students in K-5 and enjoy this interactive tool, then pass it on to another. Size matters in science, and this tool is also fun to use.

http://htwins.net/scale2/

The following website is for K-2:
The Molecularium

Supported by a grant from the National Science Foundation (NSF), the show integrates advanced scientific simulations into an immersive educational animation.

The Molecularium is part of the educational and outreach program of Rensselaer's NSF-funded Nanoscale Science and Engineering Center (NSEC) for Directed Assembly of Nanostructures.

Visit the Molecularium Web site:

http://www.molecularium.rpi.edu/

Credit: Rensselaer Polytechnic Institute

"The Strange new world of nanoscience" Winner, best short film at the Scinema Science film festival 2010.

Where and what is nano? How will it shape our future? Nanoscience is the study of phenomena and manipulation of materials at the nanoscale, where properties differ significantly from those at a larger scale. The strange world of nanoscience - it can take you into atoms and beyond the stars.

http://www.youtube.com/watch?
NR=1&feature=endscreen&v=70ba1DByUmM

Chapter 6:
NanoMission :: Learning Nanotechnology through Games

Welcome to NanoMission!

NanoMission™ is a cutting edge gaming experience, which educates players about basic concepts in nanoscience, through real world practical applications from microelectronics to targeted delivery of medicine to cancer cells.

Objective

While most young people are familiar with nanotechnology, as a futuristic technology involving miniature robots, very few have a realistic understanding of nanoscale science, or the impact on the world around them. Therefore, they are not informed about, the potential decisions they will face in the future, as nanotechnology progresses. Coupled with declining numbers of physics, chemistry and engineering students, this is a major cause for concern.

The goal is to inspire young students, about the world of nanotechnology, potentially opening their eyes to choosing it as a career, and to ensure an informed citizenry. Aimed at the gaming generations, NanoMission™ is an engaging learning experience, which educates players about basic concepts in nanoscience, through real world practical applications from microelectronics, to the medical field.

Through sponsorship, we aim to make the PC version of the game, including a 'teachers' version, which contains lesson plans and online support, **available free** to schools and colleges throughout the world. A list of our current game modules follows.

http://nanomission.org/category/about/
http://nanomission.org/category/nanomedicine-v2/
http://nanomission.org/category/nanomedicine-v1/
http://nanomission.org/category/nanoimaging/
http://nanomission.org/category/nanoscaling/

NanoMission Modules

NanoMedicine V2 Module :

This module enables a better understanding of the processes involved in creating nanomedicine. Assume the role of a biomedical scientist, aiming to cure cancer through observation and experimentation, by building nanoscopic particles and measuring their effects on the patient at the cellular level.

NanoImaging Module :

Dr Neevil has created genetically modified algae - which is manifesting in huge blooms - turning lakes red, toxic and fatal to humans and animals. The mission is to identify this micro-organism, in-order to develop a counter measure and save the world.

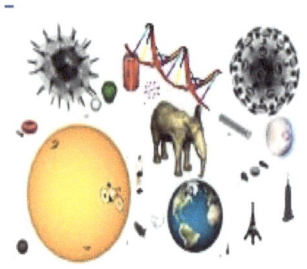

Learning Scale Module :

NanoMission scaling module enables player to visualize and understand the spatial relationships between objects at scales from the pico-meters through nano-meters, all the way up to giga-meters.

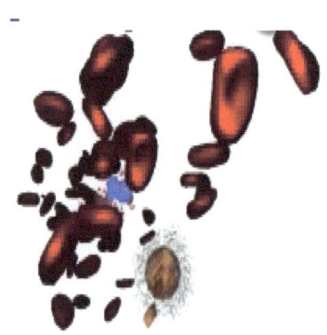

NanoMedicine V1 Module :

Join Dr Goodlove and Lisa in the demo of the first module nanomedicine. Select a suitable vehicle to deliver an anti cancer compound, and then navigate through the bloodstream to the site of the tumour, avoiding the body's natural defence mechanisms.

Nanomedicine research: Many universities are researching methods that will eliminate chemotherapy and radiation treatment, as it destroys healthy cells, along with tumours. Buckyballs and fullerenes are hollow, and can carry the medicine directly to the cancer cells.

32

Chapter 7:
"The "UVA Virtual Lab"

An NSF sponsored science education website bringing microelectronics, nanotechnology, and the underlying science to K-12 students, as well as members of the general public. It replaces math and jargon with intuitive 3D animations.

Microelectronics presentations explain how semiconductors and transistors work, and how they are fabricated in both, university labs and factories. Nanoscience presentations describe alternate forms of nanocarbon, the process of DNA self-assembly, and the inner workings of instruments used to see at the nanoscale (such as SEMs, AFMs and STMs). These pages link back to basic science presentations on electricity, magnetism and electrical circuits, including "X-ray vision" simulations, of common classroom experiments and apparatus.

Overall, the website contains over fifty presentations on micro and nanoscience, each illustrated with dozens of virtual reality animations."

The Virtual Lab experience for young students provides real laboratory experiments, that most schools cannot afford, familiarizing them with SEMs, AFMs, and STMs, normally found only in university research labs.

UVA Virtual Lab Website: www.virlab.virginia.edu

Appendix:
The Big Picture and the Future

Combining all of the videos and interactive websites, along with the games for education, creates an integrated STEM (Science,Technology,Engineering,Math), classroom where students can discuss:

1. The patterns and relationships between geometry and nanoscale science.
2. The possibilities of future technology, and how it will affect society.
3. They will also have a better understanding of the role that math plays in the world of STEM and in nature.
4. Discussions after each blended classroom video, or interactive experiment, or game should be lively and creative!

Based on the new common core standards:

Target: Elementary
Core Skills: interpretation, analysis, evaluation, explanation, inference.

Opportunities to learn science as a process of enquiry (rather than simply having "enquiry times" that are appended to an existing curriculum) has important advantages. It involves observation, imagination, and reasoning about the phenomena under study. It includes the use of tools and procedures, but in the context of authentic enquiry, these become devices that allow students to extend their everyday experiences of the world, and help them organize data in ways that provide new insights into phenomena.

Crucial questions that are not addressed by lockstep experimental exercises include the following: Where do ideas for relevant observations and experiments come from in the first place?

How do we decide what counts as relevant comparison groups?

How can nanoscale science and geometry in nature be rigorously empirical, even though they are not primarily experimental?

Definitions of what counts as "good science" change as a function of what is being studied, and current theorizing about the ideas being investigated.

One of the most important aspects of science—yet perhaps one of the least emphasized in instruction—is that science involves processes of imagination.

If students are not helped to experience this for themselves, science can seem dry and highly mechanical. Indeed, research on students' perceptions of science indicates that "they see scientific work as dull and rarely rewarding, and scientists as bearded, balding, working alone in the laboratory, isolated and lonely. Few scientists we know would remain in the field of science if it were as boring as many students believe.

Generating hypotheses worth investigating was for Einstein an extremely important part of science, where the "imagination of the possible" played a major role.

Children at play outside or with unfamiliar materials look as though they might be answering such questions as: What does this do? How does this work? What does this feel like? What can I do with it? Why did that happen?

This natural curiosity and exploration of the world around them have led some people to refer to children as "natural" scientists. Certainly these are the very types of questions that scientists pursue. Yet children are not scientists. Curiosity about how the world works, makes engaging children in science relatively easy, and their proclivity to observe and reason is a powerful tool that children bring to the science classroom.

But there is a great deal of difference between the casual observation and reasoning children engage in, and the more disciplined efforts of scientists.

37

How do we help students develop scientific ideas and ways of knowing? Introducing children to the culture of science—its types of reasoning, tools of observation and measurement, and standards of evidence, as well as the values and beliefs underlying the production of scientific knowledge—is a major instructional challenge. Children are able to take on these learning challenges successfully even in the earliest elementary grades, due to their natural curiosity.

How Students Learn: History, Mathematics, and Science in the Classroom
http://www.nap.edu/catalog/10126.html

Study shows 4 year-old preschoolers think like scientists

Scientific studies show that preschool children think like scientists. The MIT study by Laura E. Schulz and her colleague, Jessica Sommerville of University of Washington, tested 144 preschoolers. The purpose of the study was to look at whether children believe causes always produce effects. If a child believes causes produce effects deterministically, then whenever causes appear to work only some of the time, children should think some necessary cause is missing, or an inhibitory cause is present.

In one part of the study, the experimenters showed children that a switch made a toy with a metal ring light up. Half the children saw the switch work all the time; half saw the switch only lit the ring toy some of the time. They then showed the children that removing the ring stopped the toy from lighting up.

The experimenters kept the switch, gave the toy to the children and asked them to stop the toy from lighting up. If the switch always worked, children removed the ring. If the switch only worked some of the time, children could have removed the ring, but they did not – they assumed that the experimenter had some additional sneaky way of stopping the effect.

Children did something new. They picked up an object that had been hidden in the experimenter's hand (a squeezable key chain flashlight) and used that to try to stop the toy. They did not just accept that the switch might work only some of the time. They looked for an explanation.

Conclusion

This was the first study that looked at how probabilistic evidence affects children's reasoning abilities concerning unobserved causes. This research suggests that preschoolers actually have quite abstract beliefs about causal relationships. Most schools in the United States do not introduce science as a subject until grades 3-4, missing the opportunities to stimulate children who are naturally inquisitive and open.

Geometry is introduced in pre-K through 2nd grade math. By combining geometry shapes with nanoscience in these early grades, as the foundation of energy movement in nature, changes their view of the world.   Moving from separate subjects to a unified reality makes sense. The visual elements from the videos, interactive websites, and game platforms will stimulate their creative minds, and their desire to learn more about this amazing universe that we share.

Teachers will have the ability to develop learning situations in their classrooms, that stimulate ideas and questions from the visual resources that will help students 'connect the dots' that unify nature. As they enter the higher grade levels, this basic foundation of integrated learning will help them unify physics, chemistry, biology and engineering as a natural progression.

References:

Biomimmicry 3.8

http://biomimicry.net/educating/youth-education-k-12/

Graphene and solar cells:

http://news.rice.edu/2014/11/17/graphenenanotube-hybrid-benefits-flexible-solar-cells-2/

QUT leading the charge for panel-powered car

https://www.qut.edu.au/science-engineering/about/news/news?news-id=81661

How Students Learn: History, Mathematics, and Science in the Classroom

http://www.nap.edu/catalog/10126.html

Nanoscience Education, Workforce Training, and K-12 Resources
CRC Press: Light Feather, Aznar

http://www.TNTG.org

"Space buckyballs thrive, finds NASA Spitzer Telescope." Phys.org. 28 Oct 2010.

http://phys.org/news/2010-10-space-buckyballs-nasa-spitzer-telescope.html

About the Author:

Judith Light Feather
The NanoTechnology Group Inc.
A Global Nanoscience Education Consortium

It has been a long and interesting global journey since 1995, when I was invited to help shift the education paradigm by Paul Messier, from the National Learning Foundation, in Washington D.C. The Internet was just beginning and his Agile Learner project was intriguing. The network started growing rapidly and by 1996, we had one of the first websites online, "Education for the 21st Century." We submitted our proposal to the United Space Alliance, when they took over the administration for NASA, and were notified of the approval for funding in late November, 1996. In January, 1997, they did a re-organization and cancelled all the funding that had been approved. I thought it was the end of the road, but it was really just the beginning.

In 1998, I started working with the NanoComputer Dream Team and learning about this new size of science called 'nano'. We formed an educational team, and Rice University was our first member, along with NASA Johnson Space Center. Later that year, I was introduced to Dr. James Murday, who was the head of the National Nanotechnology Coordination Office, and we are still working together 16 years later. The group continued to expand and we formed The NanoTechnology Group Inc., in 2002, and co-founded CANEUS the same year.

I have travelled the globe speaking to leaders about nanoscience education for 10 years, hoping that we would start teaching our young students in K-12 someday. Virginia is the first state to include nanoscience in their Common Core Standards. NOW is the time!